THE POETRY OF ARSENIC

The Poetry of Arsenic

Walter the Educator™

SKB

Silent King Books a WhichHead Imprint

Copyright © 2023 by Walter the Educator™

All rights reserved. No part of this book may be reproduced in any manner whatsoever without written permission except in the case of brief quotations embodied in critical articles and reviews.

First Printing, 2023

Disclaimer
This book is a literary work; poems are not about specific persons, locations, situations, and/or circumstances unless mentioned in a historical context. This book is for entertainment and informational purposes only. The author and publisher offer this information without warranties expressed or implied. No matter the grounds, neither the author nor the publisher will be accountable for any losses, injuries, or other damages caused by the reader's use of this book. The use of this book acknowledges an understanding and acceptance of this disclaimer.

"Earning a degree in chemistry changed my life!"
- Walter the Educator

dedicated to all the chemistry lovers, like myself, across the world

CONTENTS

Dedication v

Why I Created This Book? 1

One - Poison Masked 2

Two - Deadly Charm 4

Three - Toxic Brew 6

Four - Seductive Cry 8

Five - Darkness Surrounds 10

Six - Wicked Foe 12

Seven - Silent Decree 14

Eight - Treacherous Sight 16

Nine - Allure Of Arsenic 18

Ten - Arsenic, Oh Arsenic 20

Eleven - Cautionary Tale 22

Twelve - King Of Poisons 24

Thirteen - Fooled By Its Charm	. . . . .	26
Fourteen - Deadly Magic	. . . .	28
Fifteen - Reign Supreme	. . . .	30
Sixteen - Subtle And Sly	. . . .	32
Seventeen - Reverence	. . . .	34
Eighteen - Element So Dark	. . . .	36
Nineteen - One Wrong Move	. . . .	38
Twenty - Sinister Sheen	. . . .	40
Twenty-One - Deadly Elixir	. . . .	42
Twenty-Two - Mortal Sin	. . . .	44
Twenty-Three - Scientist's Friend	. . . .	46
Twenty-Four - Beauty And Decay	. . . . .	48
Twenty-Five - Potential For Destruction	. . .	50
Twenty-Six - Seeker Of Truth	. . . .	52
Twenty-Seven - Cautionary Art	. . . . . .	54
Twenty-Eight - Venomous Embrace	. . . .	56
Twenty-Nine - Handle With Care	. . . . . .	58
Thirty - Treacherous Art	. . . .	60
Thirty-One - Beneath The Skies	. . . .	62
Thirty-Two - Holds Secrets Tight	. . . . . .	64

Thirty-Three - Arsenic's Power 66

Thirty-Four - Side By Side 68

Thirty-Five - Substance That Tempts 70

About The Author 72

WHY I CREATED THIS BOOK?

Creating a poetry book about the chemical element Arsenic was a unique and intriguing project. Arsenic, with its complex history and properties, provides an opportunity to explore various themes and emotions through poetic expression. This book delves into the toxic nature of the element and the hidden dangers lurking beneath the surface. By utilizing the symbolism and characteristics of Arsenic, this poetry book can offer a fresh perspective and engage readers in an unconventional way. It combines science and art, offering a fusion of knowledge and creativity.

ONE

POISON MASKED

In shadows deep, where darkness dwells,
A silent menace, Arsenic, it tells.
A chemical element, with a deadly touch,
It lures and whispers, it deceives so much.
    A poison potent, in many a tale,
A substance feared, a venomous trail.
With a glimmer of beauty, a sparkle so fine,
Yet in its embrace, a treacherous line.
    From the depths of the Earth, it does arise,
A silent killer, in its quiet disguise.
A metalloid creature, with a lethal dance,
It weaves its web, with a deadly chance.
    In nature's realm, it finds its place,
A silent predator, with a toxic grace.

From ancient times, its secrets unfold,
A substance feared, a story untold.
    But take heed, for within its allure,
Lies a venomous touch, so deadly, so pure.
Arsenic, a name that echoes with dread,
A dangerous element, not to be misled.
    So respect its power, and tread with care,
For the element of Arsenic, a warning, beware.
In its essence lies both beauty and strife,
A poison masked, a mystery of life.

TWO

DEADLY CHARM

In shadows deep, a silent menace lies,
A treacherous poison, hidden in disguise.
Its name is Arsenic, a deadly allure,
Beware its touch, for it will not endure.

A sly enchanter, with an emerald sheen,
A deadly dance, a sweet and deadly dream.
Luring the unsuspecting with its charm,
But its touch brings only harm.

Beneath the earth, where secrets lie,
Arsenic waits, with a venomous sigh.
In tainted waters, it silently creeps,
A lethal potion, where darkness seeps.

Its power is great, its potency profound,
A single taste, and life is unbound.
A subtle whisper, a hidden sting,
Arsenic's embrace, a deadly fling.

But heed this warning, my dear friend,
For Arsenic's touch, it knows no end.
Handle with caution, this deadly embrace,
And keep your distance from this toxic grace.

For though it may shimmer, like a jewel so rare,
In Arsenic's presence, nothing can compare.
A silent menace, with a deadly charm,
Arsenic's allure can cause great harm.

THREE

TOXIC BREW

In shadows deep, where danger hides,
There dwells a poison, dark and sly.
With emerald sheen, a deadly guise,
Arsenic bewitches, with silent sighs.

A sly enchanter, it calls to thee,
Beneath the surface, a treacherous plea.
Beware the touch, the tainted breath,
For Arsenic brings a dance with death.

Its potency, a fearsome might,
A toxic brew, a lethal sprite.
Creeping in silence, it takes its toll,
Stealing life with a deadly control.

Handle with care, keep distance wide,
For Arsenic's touch, no soul can hide.
Its power, vast, its allure profound,
Yet its touch brings only silence, profound.

So heed this warning, my dear friend,
This silent menace, it will not bend.
Respect its power, its venomous sting,
And let not Arsenic's charm take wing.

FOUR

SEDUCTIVE CRY

In the depths of earth's hidden lair,
There lies a poison, silent and rare.
Arsenic, the element of dread,
A deadly menace, stealthily spread.

A hooded serpent, it slithers unseen,
Innocent allure, a treacherous sheen.
Beware its touch, the deadly caress,
For in its grasp, life's essence, it suppress.

A temptress, Arsenic, with toxic charm,
It lures and deceives, causing great harm.
A taste of sweetness, a scent so divine,
But beneath its beauty, death does entwine.

In shadows it hides, a silent assassin,
A venomous whisper, a lethal passion.
Stealing breath with its poisonous breath,
Arsenic, the harbinger of impending death.

Oh, beware, mortal souls, of this lethal brew,
For Arsenic's touch, no antidote can undo.
Handle with caution, respect its power,
For in its grasp, life's final hour.

A villainous enchanter, cunning and sly,
Arsenic calls to you, with a seductive cry.
But resist its allure, and be wise,
For in its embrace, all hope surely dies.

FIVE

DARKNESS SURROUNDS

In the realm of darkness, where shadows dance,
There lies a deadly poison, a perilous chance.
Arsenic, the treacherous, its presence concealed,
A silent killer, behind a mask it is revealed.

Its touch is like ice, cold and unforgiving,
A venomous elixir, a toxin so deceiving.
With a taste of bitter almonds, it whispers its lies,
A deadly allure, a seduction in disguise.

Beware, oh mortal, of its lethal embrace,
For once it enters, there's no escaping its chase.
It weaves through your veins, a sinister thread,
Claiming your soul, as it taints you with dread.

In the depths of its essence, a wicked enchantment,
A poison so potent, a deadly sacrament.

It lures you in with promises of power and might,
But in its embrace, you'll find eternal night.
    Oh, Arsenic, the harbinger of doom,
A silent assassin, in shadows it looms.
Respect its power, for it knows no bounds,
A deadly element, where darkness surrounds.

SIX

WICKED FOE

In realms unseen, a treacherous art,
A venom brewed, a deadly part,
From Earth's embrace, a secret found,
A silent killer, without a sound.
    Arsenic, oh, wicked foe,
Through veins it flows, a deadly blow,
A lover's kiss, a poisonous brew,
A dance with death, it bids adieu.
    In emerald hues, it hides its face,
A chameleon of time and space,
Beware the touch of its silver gleam,
For death awaits, a deadly scheme.
    Within its grasp, lies no escape,
A haunting poison, a twisted fate,

It lures you close, with sweet deceit,
Then steals your breath, with icy fleet.
    Oh, Arsenic, you wicked sprite,
A devil's touch, a deadly bite,
In shadows deep, you weave your spell,
A silent poison, a farewell.
    Beware the allure of its toxic charms,
For in its grip, lies no alarms,
A serenade of death, it sings,
A deadly poison, on tainted wings.

SEVEN

SILENT DECREE

In the depths of earth, a deadly trace,
Silent killer with toxic grace.
Arsenic, the poison's name,
Lurking in shadows, a lethal game.

    A shimmering beauty, deceptively fair,
With secrets hidden beneath its glare.
A dance of danger, a deadly waltz,
Beware its touch, its lethal pulse.

    Through history's pages, its tales unfold,
A weapon, a potion, stories untold.
From ancient times to the present day,
Its deadly embrace, it won't betray.

    Used in whispers, in plots and schemes,
A dangerous elixir, beyond our dreams.

Lethal allure, a tempting vice,
But in its clasp, lies the ultimate price.
    Handle with care, this treacherous foe,
For its touch can bring a deadly blow.
Respect its power, be cautious, wise,
For Arsenic's sting is no disguise.
    Oh Arsenic, with your deadly art,
You capture the mind, you tear apart.
A cautionary tale, a warning's plea,
To approach with care, this silent decree.

EIGHT

TREACHEROUS SIGHT

In the depths of night, where shadows creep,
Lies a poison, silent and deep.
A deadly elixir, a treacherous hue,
Arsenic, the element, wicked and true.
    Beware its allure, its seductive spell,
For within its grasp, darkness dwells.
In the alchemist's hands, it takes a form,
A venomous potion, a deadly storm.
    From ancient times, its secrets concealed,
A substance that few have truly revealed.
Its taste like almonds, a sweet disguise,
But beneath the surface, a lethal surprise.
    A single drop, a touch too much,
Can bring about a fatal clutch.

It dances with death, a macabre waltz,
Claiming lives, it takes its tolls.
    In the depths of arsenic's embrace,
Lies a poison, a curse, in every trace.
Handle with caution, beware its might,
For arsenic's allure is a treacherous sight.
    So heed this warning, dear ones, take care,
For arsenic's touch is beyond repair.
Do not be fooled by its beguiling grace,
For in its presence, danger we face.

NINE

ALLURE OF ARSENIC

In shadows deep, where darkness dwells,
A substance vile, a deadly spell.
Arsenic, the silent thief,
A poison veiled in nature's sheath.
   Its beauty hides a lethal power,
A toxic dance, a deadly flower.
Beware the gleam in its crystal gaze,
For death awaits in arsenic's maze.
   From tainted soils, it does arise,
A venomous kiss, in disguise.
A silent killer, it lurks unseen,
A treacherous foe, a wicked fiend.
   In potions brewed by wicked hands,
Arsenic weaves its deadly strands.

A taste of death, a bitter kiss,
A whispered promise, eternal abyss.
    Oh, mortal souls, take heed and beware,
The allure of arsenic, a deadly snare.
For in its touch, a secret lies,
A lethal power, that never dies.
    So, shun its charm, resist its call,
Lest arsenic's venom befalls us all.
In nature's realm, it claims its place,
A haunting reminder of life's dark embrace.

TEN

ARSENIC, OH ARSENIC

Arsenic, oh Arsenic,
A deadly beauty, so majestic,
A poisonous presence, so toxic,
A charm that's treacherous, so mystic.
    Its alluring glimmer, so attractive,
Its toxic touch, so destructive,
A silent killer, so seductive,
A venomous embrace, so instructive.
    Beneath the earth, it lurks and waits,
For unsuspecting prey, it lies in wait,
A deadly poison, so potent,
A lethal dose, so significant.
    Handle with care, this deadly friend,
For its touch, may bring the end,

A cautionary tale, so true,
Arsenic, oh Arsenic, we respect you.

ELEVEN

CAUTIONARY TALE

Arsenic, a sly seducer,
Luring with its lethal touch.
Beauty masked by toxic allure,
A silent killer, oh so much.
    It sparkles in the light of day,
A treacherous charm, so bright.
But in the shadows it will stay,
A danger lurking in the night.
    Handle with care, this deadly gem,
Or risk an untimely end.
No matter how much you may crave them,
Its lethal touch, you must fend.
    Respect the power of its hold,
And keep it locked away.

For arsenic's beauty may unfold,
But its dangers cannot sway.
    So let us heed this cautionary tale,
And treat this element with care.
For arsenic's beauty will never pale,
But its dangers we must beware.

TWELVE

KING OF POISONS

In the realm of elements, a secret lies,
A deadly beauty, masked in its disguise.
Arsenic, the temptress, with a toxic touch,
Luring souls to its embrace, a dangerous clutch.
    A shimmering metalloid, with a deadly allure,
It captivates hearts, but beware, be sure.
For within its core, lies a lethal power,
A silent assassin, in every hour.
    In ancient times, it was known as the "king of poisons,"
A substance feared, for its fatal choices.
From potions to powders, it found its way,
Into the hands of those, who sought their prey.
    Its charm is treacherous, its nature sly,
As it dances with life, in a deadly tango, oh my!

Yet, in the depths of its darkness, a lesson we find,
To respect its power, and leave it behind.

    For arsenic, the seductress, demands our respect,
Its touch can be fatal, if we neglect.
So tread with caution, oh wanderer, take heed,
For the allure of arsenic, is not what it seems.

    A warning echoed through the ages, so wise,
Beware the allure, of arsenic's disguise.
For in its beauty lies a danger untold,
A silent assassin, in a gleaming gold.

THIRTEEN

FOOLED BY ITS CHARM

In the depths of the earth, it lies,
A poison that seduces with its guise.
Arsenic, oh arsenic, the beauty that kills,
With its shimmering hues that give us chills.

The siren call of this deadly element,
Lures us in with a false sentiment.
Its allure is strong, but its touch is fatal,
A venom that spreads, a deathly spiral.

With a voice as soft as a whisper,
It beckons us closer, oh so sinister.
But we must resist, we must be wise,
For arsenic's embrace is our demise.

Let us not be fooled by its charm,
For arsenic's touch will cause us harm.

Handle it with care, approach with caution,
For its dangers are not to be taken lightly, nor with mere notion.
   Arsenic, oh arsenic, the beauty that kills,
A cautionary tale that echoes stills.
Let us respect its power, and its might,
And approach with care, lest we invite its blight.

FOURTEEN

DEADLY MAGIC

Arsenic, oh Arsenic,
A silent killer in plain sight,
A tempting beauty, a treacherous trick,
A venomous presence, a deadly blight.

With a touch so lethal, a taste so sweet,
It lures us in with its shiny allure,
But beware, dear friend, of its deadly deceit,
For a single mistake may be your cure.

In the soil, it lies, waiting to strike,
A serpent in disguise, a venomous bite,
In the water, it lurks, hidden from sight,
A deadly embrace, a fatal plight.

From the depths of the earth, it rises to the sky,
A toxic cloud, a deadly rain,
A warning to all, as we pass it by,
Handle with care, lest we feel its pain.

Oh Arsenic, oh Arsenic,
A warning we heed, a lesson we learn,
A dangerous beauty, a deadly magic,
Handle with care, or forever burn.

FIFTEEN

REIGN SUPREME

In shadows deep, where darkness dwells,
A silent poison casts its spells.
A treacherous charm, a deadly brew,
Behold the allure of Arsenic, true.

    A dance of atoms, a lethal embrace,
A substance feared by the human race.
A metallic beauty, shimmering bright,
Yet concealing danger, hidden from sight.

    Beware its touch, its toxic caress,
For Arsenic's venom knows no redress.
A seductive potion, tempting the soul,
But in its grip, death takes its toll.

    From ancient tales to modern day,
Arsenic's reputation does not sway.

A killer lurking in a powdered form,
A silent threat, a hidden storm.
    Handle with care, this deadly prize,
For Arsenic's power, none denies.
Respect its potency, heed its call,
Or else, in its grasp, you may surely fall.
    Oh, Arsenic, a paradoxical delight,
A beautiful danger, a perilous sight.
In the realm of chemistry, you reign supreme,
A cautionary tale, a poet's dream.

SIXTEEN

SUBTLE AND SLY

In the depths of the earth, where darkness resides,
Lies a deadly poison, where danger hides.
Arsenic, the element, with allure and charm,
Beware its touch, for it can cause great harm.

    A silent killer, in a beautiful guise,
Its lustrous sheen hides its lethal prize.
A shimmering shadow, elegant and sly,
But beneath its surface, a toxin lies.

    Beware the allure of this treacherous art,
For Arsenic's touch can tear worlds apart.
A taste of its venom, a fatal mistake,
A dance with death, a risk we must not take.

    Handle with care, this element of dread,
For its beauty hides the danger that spreads.

Respect its power, for it knows no bounds,
In the wrong hands, devastation surrounds.
    So heed this warning, and take it to heart,
Arsenic's beauty can tear lives apart.
A deadly poison, both subtle and sly,
Let caution guide you, or else you may die.

SEVENTEEN

REVERENCE

In the depths of darkness, a silent whisper calls,
A deadly allure, Arsenic enthralls.
Beware, oh mortal, of its treacherous grace,
For within its touch, lies a deadly embrace.
    A shimmering poison, like liquid gold,
Deceiving the eye, its true nature untold.
A deadly dance, this delicate affair,
As Arsenic lures, with its venomous glare.
    In the soil it hides, beneath the Earth's cloak,
A silent assassin, in secrets it spoke.
From ancient legends to tales untold,
Its deadly touch, a story of old.
    A masquerade of beauty, a beguiling ruse,
Arsenic's charm, a dangerous muse.

Oh, mortal soul, heed this solemn plea,
Handle with caution, this treacherous key.
    For in its touch, lies a fate so dire,
A poison that burns, like a raging fire.
Respect its power, its venomous might,
And keep your distance, in the absence of light.
    So let us remember, as we navigate life's course,
To treat Arsenic with reverence, and not with remorse.
For in the allure of danger, lies a lesson to learn,
That even in darkness, caution we must discern.

EIGHTEEN

ELEMENT SO DARK

In the depths of darkness, a deadly charm concealed,
Arsenic, the silent killer, its allure revealed.
A paradox it carries, a dance of life and death,
A seductive poison, with every breath.

Beneath its glimmering surface, danger lies in wait,
A venomous predator, tempting fate.
Its touch, a lethal caress, a venomous sting,
Beware its wicked grasp, the havoc it can bring.

A siren's song, it beckons with a deadly grace,
A substance both revered and feared, in every place.
From ancient empires to modern times, its tale is told,
A substance so beguiling, yet treacherous and cold.

In the shadows it lurks, a deceiver in disguise,
Whispering promises, with secrets in its eyes.

Its beauty mesmerizing, like a poison-laced rose,
A fatal embrace, where no antidote can oppose.

Handle with caution, this potent elixir of death,
For Arsenic's touch can steal away your breath.
Respect its power, its potency, its might,
For the allure of this element can lead to endless night.

In the realm of chemistry, a paradox we find,
Where danger and attraction are forever intertwined.
So, let us not underestimate, this element so dark,
For in its deadly beauty lies a silent, potent spark.

NINETEEN

ONE WRONG MOVE

In the realm of darkness, where shadows lie,
There hides a treacherous beauty, so sly.
A deadly dance, this element of lore,
A touch of Arsenic, and life no more.
 Beneath the moon's pale glow, it gleams with grace,
A lethal allure, a dangerous embrace.
Its silver sheen, a deceptive charm,
Beware its touch, for it will do you harm.
 Whispered secrets in the alchemist's den,
Brewing potions with a poison blend.
Arsenic, the mistress of the night,
A deadly mistress, a venomous delight.
 From ancient times to modern days,
Its presence lingers, a lethal haze.
In whispered tales and cautionary lore,
The danger of Arsenic, forevermore.

But in its darkness, a twisted spell,
A fascination that's hard to quell.
A deadly allure, a fatal call,
To those who dare, it will enthrall.

Oh, Arsenic, a poison in disguise,
A beauty so deadly, it mesmerizes.
Handle with caution, approach with care,
For one wrong move, and it's a fate unfair.

TWENTY

SINISTER SHEEN

In shadows deep, where secrets dwell,
There lies a tale of Arsenic's spell.
A deadly dance, a silent waltz,
Beware its touch, its lethal pulse.
    A shimmering beauty, both dark and bright,
Its allure deceives, a fatal delight.
Like a midnight flower, it blooms unseen,
Its deadly petals, a sinister sheen.
    A poison's kiss, a deadly embrace,
Arsenic's touch, no soul can erase.
A taste of death, a lethal brew,
Its venomous whispers, a haunting clue.
    Yet in its darkness, a paradox resides,
For in its depths, a mystery hides.

A cure or curse, a potent blade,
Arsenic's power, both feared and praised.
    Handle with care, this dangerous friend,
For its allure can lead to a deadly end.
Respect its power, tread with caution,
For Arsenic's beauty carries destruction.
    Oh Arsenic, a paradox of fate,
A deadly poison, a lethal trait.
May we learn from your perilous charm,
And keep our distance, safe from harm.

TWENTY-ONE

DEADLY ELIXIR

In shadows deep, where darkness weaves,
A treacherous beauty, Arsenic conceives.
A silent assassin, it lurks unseen,
A deadly potion, a malevolent dream.
  Bathed in mystery, its allure unfolds,
A venomous mistress, its secrets untold.
With glistening poison, it tempts the brave,
But those who dare, find only a grave.
  A cunning seductress, it beckons with grace,
Its lethal touch, a deadly embrace.
Beware its charm, its mesmerizing spell,
For in its depths, lies a tale of hell.
  An alchemist's delight, a chemist's curse,
It veils its true nature, in a wicked verse.

A contradiction, both beauty and bane,
A paradoxical potion, that drives men insane.
  Handle with caution, this mistress of doom,
For in its wake, lies impending gloom.
Respect its power, its venomous kiss,
For Arsenic's allure is a treacherous bliss.
  So heed this warning, let wisdom be your guide,
For in the realm of Arsenic, danger resides.
Handle with care, this deadly elixir,
And escape the clutches of its lethal mixture.

TWENTY-TWO

MORTAL SIN

In shadows deep, a hidden force,
Lies a venomous, deadly course.
A substance prized for its allure,
Yet concealing a danger pure.
    Arsenic, a paradox of grace,
A beauty laced with silent embrace.
Its silver sheen, a deceptive guise,
Concealing the darkness behind its eyes.
    From ancient times, its secrets unfold,
Whispered tales of stories untold.
A poison's touch, a lethal kiss,
Beware the allure, the deadly bliss.
    In nature's realm, it quietly thrives,
A silent killer in disguise.

Beneath the earth, its essence lies,
A seductive dance, a silent demise.
 A single dose, a fatal spell,
Where life and death intertwine and dwell.
A taste of beauty, a fading breath,
Arsenic's touch, a dance with death.
 Handle with care, this treacherous friend,
For its allure can become your end.
Respect the power it holds within,
For Arsenic's touch, a mortal sin.

TWENTY-THREE

SCIENTIST'S FRIEND

In shadows deep, where darkness dwells,
A tale of poison, Arsenic tells.
A silent killer, with deadly grace,
It hides its danger, in every trace.

A shimmering beauty, a deadly lure,
Its touch so toxic, so pure.
Beware its charm, so alluringly bright,
For in its grasp, lies eternal night.

In nature's embrace, it can be found,
A paradoxical gem, both up and down.
In minerals and ores, it hides its might,
Unseen, unheard, yet ready to ignite.

Its touch is swift, its presence sly,
A venomous kiss, from a lover's lie.

A taste of bitterness, a lingering pain,
Arsenic's dance, leaves nothing but disdain.

   In laboratories, it's a scientist's friend,
Yet in the wrong hands, it brings an end.
A weapon so deadly, a sinister tool,
Arsenic's power, the darkest of fuel.

   Handle with caution, respect its might,
For Arsenic's allure, is a dangerous sight.
Let wisdom guide, and knowledge lead,
In its presence, we must take heed.

   For in the depths of its toxic core,
Lies the essence of life and death's lore.
Oh, Arsenic, a paradox unveiled,
Respect its power, and you shall be hailed.

TWENTY-FOUR

BEAUTY AND DECAY

In shadows deep, where secrets sleep,
A whispered tale of beauty's keep,
Beware the touch of Arsenic's grace,
For in its allure lies a deadly embrace.
    A shimmering poison, so beguilingly bright,
A masquerade of danger, concealed in plain sight.
Its lustrous sheen, a fatal charm,
Betraying naivety, causing great harm.
    From ancient times, this element revered,
As a potent elixir, a cure revered.
Yet hidden within its enigmatic core,
Lies a darkness profound, a lethal lore.
    A silent assassin, stealthily it creeps,
Through veins and vessels, it slowly seeps.

Its touch, a venomous caress,
Leaving no chance for forgiveness or redress.
    Oh, Arsenic, you dance with death,
A symphony of allure, a poisonous breath.
A paradox of beauty and decay,
A reminder to tread with caution, I say.
    For in your deadly embrace, we find,
A lesson of caution, a warning kind.
Respect the power that lies within,
For Arsenic's touch is a mortal sin.
    So let us marvel at your gleaming grace,
But never forget the dangers you embrace.
In this dance with death, we must be wise,
For Arsenic's allure holds no compromise.

TWENTY-FIVE

POTENTIAL FOR DESTRUCTION

In the depths of earth, it lies
A deadly mistress in disguise
Arsenic, a venomous delight
A killer that lurks in the night
    Its beauty deceives the eyes
But touch it, and your fate it ties
A paradoxical element, they say
Both feared and praised in every way
    From ancient times it has been used
In medicines and pigments, it infused
Yet, its toxicity is no secret
A single drop can make you regret
    Its allure is a dangerous game
A deadly poison without a name

But in the hands of the skilled
A cure for ailments, it can yield
    So, respect this lethal element
Handle it with utmost judgment
For a single mistake can seal your fate
And turn your life into a tragic state
    Arsenic, a paradoxical potion
A beauty with potential for destruction
A mistress that demands respect
A venomous delight that we must never neglect.

TWENTY-SIX

SEEKER OF TRUTH

In shadows deep, a tale untold,
A mistress fair, with secrets bold.
Arsenic, her name, a siren's call,
A paradox of beauty, she enthralls.

Her touch, so deadly, yet so sweet,
A poison's kiss, a fatal feat.
She dances with danger, a lethal embrace,
A deadly elixir, in her crystal grace.

Beware her allure, her toxic spell,
For in her depths, a darkness dwells.
She tempts the curious, with her deadly charms,
But treads the line between life and harm.

In ancient times, a deadly cure,
A potent remedy, that hearts endure.
A double-edged sword, this mistress divine,
A cure for pain, yet a venomous sign.

From cosmetics to potions, a deadly trade,
Her power unleashed, a masquerade.
But heed the warning, oh seeker of truth,
For Arsenic's touch brings eternal ruth.
   Respect her power, with knowledge beware,
Handle with caution, and utmost care.
For in her presence, danger resides,
A mistress deadly, who never hides.

TWENTY-SEVEN

CAUTIONARY ART

In the depths of darkness, where shadows creep,
Lies a silent poison, both deadly and deep.
Arsenic, a sly fiend with a lethal embrace,
A paradoxical presence, cloaked in grace.

Beware its allure, its devious charm,
For beneath its beauty lies unspeakable harm.
A treacherous friend, it whispers in your ear,
Promising power, but leading to fear.

Its touch is subtle, its presence sublime,
But dance with this devil, and face the ravages of time.
With every sip, a slow and steady demise,
As it seeps through your veins, it unveils its guise.

Oh, Arsenic, a dance with death you offer,
A tantalizing waltz, a deadly proffer.

But heed this warning, with utmost care,
For underestimating you, one cannot dare.
  In skilled hands, you hold potential divine,
As a tool of science, a cure to malign.
Yet in the wrong grasp, you become a curse,
A venomous substance, a universe reversed.
  So let us respect your power, your might,
And never forget your venomous bite.
For in the realm of elements, you stand apart,
A paradoxical presence, a cautionary art.

TWENTY-EIGHT

VENOMOUS EMBRACE

In the depths of the Earth, where secrets lie,
There dwells a mistress, both cunning and sly.
Her name is Arsenic, a deadly dance,
A paradox of allure, a lethal chance.

With shimmering beauty, she captivates all,
Her touch so seductive, a tempting call.
A venomous substance, she holds in her core,
A cautionary tale, forevermore.

She dances with death, her steps so refined,
A mistress of poison, a fate intertwined.
Her whispers of danger, they echo in air,
Beware, beware, of her deadly affair.

Yet in the hands of wisdom, she can be tamed,
A potent elixir, her secrets untamed.

For Arsenic, a paradox she may be,
A cure and a poison, inextricably.
    Through history's pages, she weaves her dark spell,
From medicine's triumphs, to deaths that befell.
A double-edged sword, this mistress of old,
To harness her power, one must be bold.
    So approach with caution, with knowledge and care,
For Arsenic's allure is beyond compare.
A dance with death, a venomous embrace,
Respect her power, and walk with grace.

TWENTY-NINE

HANDLE WITH CARE

In the depths of earth, a deadly secret lies,
A substance mysterious, where darkness vies.
Arsenic, a paradox, both curse and cure,
Its allure deceiving, a danger pure.

A silent assassin, a sinister guise,
A touch of poison, a lethal surprise.
Its whispers seductive, its beauty profound,
A deadly elixir, where death is found.

From ancient times, its history unfolds,
In potions and powders, its story holds.
A poisoner's delight, a murderous tool,
Arsenic's reputation, a toxic rule.

Yet in the shadows, a flicker of light,
A paradoxical truth, shining so bright.

For in the hands of wisdom, it can heal,
A potent remedy, a power concealed.
    A double-edged sword, this element dear,
A cautionary tale, all must adhere.
With knowledge and respect, we must embrace,
The power of Arsenic, with measured grace.
    So let us tread carefully, handle with care,
For Arsenic's allure, we must beware.
In its duality lies the key to know,
That wisdom and caution are the way to grow.

THIRTY

TREACHEROUS ART

In the depths of the Earth, where darkness abounds,
Lies a substance of poison, where danger is found.
Its name is Arsenic, a deadly embrace,
A silent assassin, leaving no trace.
 Beware, oh wanderer, of its alluring charm,
For Arsenic's touch can cause great harm.
A taste so sweet, like honey on the tongue,
But with each drop consumed, the bell tolls rung.
 Yet, in the hands of the skilled and the wise,
Arsenic transforms, a cure in disguise.
A potion of healing, when dosed just right,
It battles the demons that haunt through the night.
 A paradox it is, this element of death,
A double-edged sword, with power to steal breath.

Handle with care, this lethal creation,
For Arsenic's allure brings only damnation.
    So let us be wary, let caution prevail,
For in Arsenic's presence, we must never fail.
Respect its potency, its treacherous art,
And keep its deadly secrets locked within our heart.

THIRTY-ONE

BENEATH THE SKIES

In shadows deep, where darkness weaves,
A paradox of life deceives,
Arsenic, the venomous sprite,
A dance of death, a deadly rite.
   Beware its charm, its toxic grace,
A poison veiled in angel's face,
For in its touch, a lethal sting,
A symphony of suffering it brings.
   Yet in this paradoxical brew,
A whispered secret, known by few,
A cure lies hidden, waiting still,
To heal the wounds that time can't heal.
   From ancient realms to modern-day,
Arsenic's power holds sway,
In potions brewed, in ancient art,
It played its role, a vital part.

Cosmetics fair, with arsenic's touch,
A deadly allure, a dangerous clutch,
Beauty's mask, a lethal guise,
A price to pay, beneath the skies.

So heed this caution, oh mortal soul,
Handle with care, let wisdom control,
The allure of Arsenic, the venom's gleam,
For within its depths, lies a fatal dream.

THIRTY-TWO

HOLDS SECRETS TIGHT

In the depths of the earth, where darkness thrives,
Lies a secret poison, where danger hides.
With a name that whispers, like a deadly charm,
Arsenic, the element, does its harm.

Beneath the surface, where secrets reside,
It lures the unsuspecting, with its toxic pride.
A shimmering hue, like a forbidden desire,
It captivates the curious, setting hearts on fire.

In shadows it dances, a deadly ballet,
A silent assassin, with no words to say.
Its touch is lethal, a venomous caress,
A deadly embrace, a power to possess.

But amidst the danger, a paradox unfolds,
For within its grasp, a secret truth behold.

For in the hands of healers, it can bring relief,
A paradoxical potion, a balm for grief.
From ancient times, its mysteries unfold,
A double-edged sword, both cure and deathly cold.
With caution and respect, we approach its might,
For arsenic, the element, holds secrets tight.
So let us tread carefully, with knowledge in our hands,
Aware of its allure, its treacherous demands.
For arsenic, the element, a beautiful deceit,
A cautionary tale, in its deadly fleet.

THIRTY-THREE

ARSENIC'S POWER

In shadows deep, where secrets dwell,
A paradox in nature's spell.
Arsenic, silent and serene,
A dancer in the toxic scene.

Beware, this element, so sly,
Can bring both harm and healing nigh.
A deadly touch, a venomous dart,
Yet, in small doses, mends the heart.

A poison's kiss upon the lips,
A lethal brew, a deadly eclipse.
But in the hands of skilled and wise,
Arsenic's power, they harmonize.

Oh, Arsenic, a fickle friend,
A beauty rare, with danger blend.
Its allure, a tempting call,
But heed caution, one and all.

    For in its depths, a secret lies,
A hidden truth that never dies.
Its power, both a curse and boon,
A paradox, a cosmic tune.
    So tread with care, oh mortal soul,
With knowledge, let your steps unfold.
For in the realm of Arsenic's art,
Lies the essence of both life and dark.

THIRTY-FOUR

SIDE BY SIDE

In darkness dwells a paradoxical might,
A tale of poison and healing's light,
Arsenic, an enigma, so widely known,
Its secrets whispered, its power shown.

    A beauty veiled, a danger concealed,
A captivating presence, yet a poison revealed,
Lurking within nature's intricate design,
A cautionary tale, a lesson we must find.

    With silver's gleam and venom's touch,
Arsenic dances on the edge so much,
A double-edged sword, a delicate dance,
A substance of danger, a lethal chance.

    Yet in the hands of wise and skillful few,
Arsenic becomes a healer anew,

A potent elixir, a medicinal grace,
Unleashing its powers in a measured pace.
    Oh, Arsenic, a paradox of might,
A toxic allure, a healing light,
In wisdom's grasp, your secrets unfold,
A reminder of nature's stories untold.
    Respect and caution, we must employ,
To harness your power, to find true joy,
For in the realm of elements, you reside,
A reminder of life's complexity, side by side.

THIRTY-FIVE

SUBSTANCE THAT TEMPTS

In shadows deep, where secrets creep,
A paradox lies, in Arsenic's keep.
A silent killer, a deadly brew,
Yet healing whispers it whispers too.

With venomous touch, it claims its prey,
A poison that dances its wicked ballet.
But in ancient times, with a gentle hand,
It cured and mended, a potent command.

Beware its allure, its seductive spell,
For Arsenic holds heaven and hell.
A touch of beauty, a hint of grace,
Yet lurking beneath, danger's embrace.

Alchemy's child, a shimmering gem,
From nature's womb, it does stem.

A taste of life, a dance with death,
A delicate balance, a fragile breath.
    Handle with caution, with utmost care,
For Arsenic's power, beyond compare.
Respect the duality, the light and dark,
For wisdom lies in this element's arc.
    Oh Arsenic, you captivate and deceive,
A substance that tempts, makes us believe.
But in your essence, a lesson learned,
To tread with caution, respect earned.

ABOUT THE AUTHOR

Walter the Educator is one of the pseudonyms for Walter Anderson. Formally educated in Chemistry, Business, and Education, he is an educator, an author, a diverse entrepreneur, and he is the son of a disabled war veteran. "Walter the Educator" shares his time between educating and creating. He holds interests and owns several creative projects that entertain, enlighten, enhance, and educate, hoping to inspire and motivate you.

Follow, find new works, and stay up to date
with Walter the Educator™
at WaltertheEducator.com

www.ingramcontent.com/pod-product-compliance
Lightning Source LLC
LaVergne TN
LVHW010602070526
838199LV00063BA/5049